# CHEMISTRY AND THE ENVIRONMENT
# VOLUME 1

## THINGS YOU SHOULD KNOW
## (QUESTIONS AND ANSWERS)

By Rumi Michael Leigh

## Introduction

I would like to thank and congratulate you for purchasing this book, "*Chemistry and the environment, things you should know (questions and answers)*".

This book will help you understand, revise and have a good general knowledge and keywords of chemistry and the environment.

Thanks again for purchasing this book, I hope you enjoy it!

Thanks again for purchasing this book, I hope you enjoy it!

## Questions: Part 1

1) What is a chemical process?
2) Give an example of a chemical process.
3) What is a physical process?
4) Give an example of a physical process.
5) What is sublimation?
6) What is vaporization?
7) What is evaporation?
8) What are the 2 types of vaporisation?
9) Do liquids have a fixed volume?
10) Do solids have a fixed volume?

# Answers: Part 1

1) A chemical process is when there is transformation of matter.
2) An example is burning wood.
3) A physical process is when there is no transformation of matter.
4) The melting of ice.
5) It is the change from a solid state to a gaseous state without going through a liquid state.
6) It is the change from a liquid state to a gaseous state.
7) It is the change from a liquid state to a gaseous state (It depends little on the increase in temperature).
8) Evaporation and boiling.
9) Yes.
10) Yes.

## Questions: Part 2

1) What are the three main states of matter?
2) Why are the particles of a liquid united?
3) What does temperature depend on?
4) Are gases compressible?
5) Do gases have a fixed form?
6) Do gases have a fixed volume?
7) What is the opposite reaction of vaporization?
8) What is an amorphous solid?
9) What is the reverse reaction of fusion?
10) What is the reverse reaction of sublimation?

## Answers: Part 2

1) Solid, liquid and gas.
2) They are united because of their cohesive forces.
3) It depends on the agitation of its particles.
4) Yes.
5) No.
6) No.
7) Liquefaction.
8) It is a solid without a regular structure.
9) Solidification.
10) Reverse sublimation (Deposition).

## Questions: Part 3

1) What part of a liquid does evaporation occur?
2) Does evaporation occur at a certain temperature?
3) What is heat?
4) What is fusion?
5) What is ebullition?
6) What is a molecule?
7) What is an endothermic body?
8) What is an exothermic body?
9) What is the absolute zero?
10) What is the connection between Kelvin and degree Celsius?

## Answers: Part 3

1) Any part of the liquid.
2) No, it takes place at any temperature.
3) It is thermal energy given or removed from a body.
4) Fusion is the move from a solid state to a liquid state.
5) Ebullition is the move from a solid state to the gaseous state at a precise temperature.
6) It is a combination of at least two atoms.
7) It is a body that absorbs heat.
8) It is a body that gives off heat.
9) This is the zero Kelvin, the lowest temperature.
10) 0 Kelvin equals -273.15 degrees Celsius.

## Questions: Part 4

1) What part of the liquid does evaporation occur?
2) What is the relationship between pressure and temperature?
3) What is matter?
4) What is photolysis?
5) What is thermolysis?
6) What is the relationship between altitude and pressure?
7) What is a pure substance?
8) What are mixtures?
9) What is the balance between a homogeneous and heterogeneous mixture?
10) What is electrolysis?

## Answers: Part 4

1) On the surface of the liquid.
2) As pressure increases, temperature also increases.
3) Matter is anything that has a mass and occupies space.
4) It is separation by light.
5) It is separation by heat.
6) Pressure decreases as altitude increases.
7) It is a substance with the same atoms.
8) It is the combination of a minimum of 2 different pure substances.
9) It is a colloidal mixture.
10) Electrolysis is separation by electricity.

## Questions: Part 5

1) What charge does a proton have?
2) What charge does a neutron have?
3) What charge does an electron have?
4) Name the 3 types of intramolecular bonds.
5) What are isotopes?
6) Explain a metal bond.
7) Explain a covalent bond.
8) Explain an ionic bond.
9) What is hydrolysis?
10) What constitutes an atom?

## Answers: Part 5

1) A positive charge.
2) A neutron does not have a charge.
3) A negative charge.
4) Ionic, covalent and metallic bonding.
5) They are elements with the same number of protons but variable neutrons.
6) It is a connection between a metal element and another metal element.
7) It is a bond between a non-metal element and a non-metal element.
8) It is a connection between a metal element and a non-metal element.
9) It is separation by water.
10) An atom consists of proton, neutron and electron.

## Questions: Part 6

1) What is the product of a covalent bond?
2) What is the product of an ionic bond?
3) What is ionization?
4) What is an ion?
5) What do the different electronic layers correspond to?
6) What is an atomic number?
7) What is an elementary charge?
8) What is a mass number?
9) What is the number of nucleus?

**Answers: Part 6**

1) A molecule.
2) A salt.
3) It is the energy needed to remove an electron from an atom.
4) It is an atom that carries a charge.
5) They correspond to different energy levels.
6) It is the number of protons.
7) It is the smallest charge.
8) It is the number of nucleons.
9) It is the number of protons and the number of neutrons.

## Questions: Part 7

1) What does the word "ate" mean?
2) What does the word "ite" mean?
3) Give an example of a solute.
4) Give an example of a solvent.
5) Why is water considered a universal solvent?
6) What does it mean that an atom is looking to fill up its last electronic layer?
7) Does salt conduct electricity?
8) How can salt conduct electricity?
9) What are the components of a solution?
10) What is the condition for a substance to conduct electricity?

## Answers: Part 7

1) It means more oxygenated.
2) It means less oxygenated.
3) Sugar.
4) Water.
5) It is because water dissolves a large number of substances.
6) This means that the atom tries to have 8 electrons on its last peripheral layer.
7) No.
8) It must be in a state of solution.
9) A solvent and a solute.
10) The substance must have free charges. Charges that can move freely.

## Questions: Part 8

1) What is the force with which one atom attracts the electrons of another neighbouring atom?
2) What is the force that allows water not to overflow out of a glass (filled with water)?
3) What is intermolecular binding?
4) Which cations and anions attract each other?
5) What is the name of the force that binds the rare gases?
6) Which has a higher density, water or ice?
7) What is an exothermic reaction?
8) What is an endothermic reaction?
9) What influences the speed of a reaction?
10) What is used to measure the amount of heat during a reaction?

## Answers: Part 8

1) Electronegativity.
2) The surface tension.
3) It is the force that creates the link between the molecules.
4) The Coulomb force.
5) The Van der Waals force.
6) The density of water.
7) It is a reaction that gives off heat.
8) It is a reaction that absorbs heat.
9) Temperature, contact area, concentration and pressure.
10) A calorimeter.

## Questions: Part 9

1) Give an example of a table sugar.
2) What is the best known or most known sugar?
3) What is a carbohydrate?
4) What is organic chemistry?
5) What are the 2 great compositions of dry air?
6) Give their percentages.
7) What is inorganic chemistry?
8) What are hydrocarbons?
9) What are aliphatic hydrocarbons?
10) What are alicyclic hydrocarbons?

**Answers: Part 9**

1) Sucrose.
2) Glucose.
3) It is a set of sugar and polymers.
4) It is the chemistry of carbon. The chemistry of living organisms.
5) Nitrogen and oxygen.
6) About 79% of nitrogen and 21% of oxygen.
7) It is the chemistry of minerals such as water, etc.
8) Carbon and hydrogen.
9) Hydrocarbons that consist of open carbon chains.
10) Hydrocarbons that consist of closed organic chains.

## Questions: Part 10

1) What does the reaction between an acid and a base give?
2) What is effervescence?
3) What is an acid?
4) What is a base?
5) What is a strong acid?
6) What is a weak acid?
7) What constitutes alcohol?
8) What are aromatic hydrocarbons?
9) What is a reduction reaction?
10) What is an oxidation reaction?

## Answers: Part 10

1) It gives water and salt.
2) This is when acids react with hydrogen carbonates or carbonates to form carbon dioxide and water.
3) An acid is a substance that dissociates in aqueous solution to produce hydrogen ions (H+).
4) A base is a substance that dissociates in an aqueous solution to produce hydroxide ions (OH-).
5) An acid is strong when all its molecules are completely ionized in water.
6) An acid is weak when its molecules partially ionize in water.
7) Alcohol consists of a hydroxyl group attached to a carbon.
8) These are hydrocarbons that consist of benzene rings.
9) It is a chemical reaction during which there is a reduction in the number of oxidation of an element.

10) It is a chemical reaction during which there is an increase in the number of oxidation of an element.

## Questions: Part 11

1) Is an oxidant an electron donor?
2) Is a reducer an electron donor?
3) What is an ampholyte?
4) Give an example of a known ampholyte.
5) What is the pH of the blood?
6) What is neutralization of a solution?
7) What are the values of a pH scale?
8) Name some catalysts.
9) What influences chemical equilibrium?
10) Does a catalyst decrease or raise the activation energy during a chemical reaction?

## Answers: Part 11

1) No, it is an electron acceptor.
2) Yes, it is an electron donor.
3) It is a molecule that can behave either as an acid or a base.
4) Water.
5) From 7.35 to 7.45.
6) It is to remove the acidity or the basicity of a solution.
7) From 0 to 14.
8) Acids or bases, enzymes and metals such as iron, Pd.
9) Concentration, temperature and pressure.
10) A catalyst decreases the activation energy.

# Questions: Part 12

1) Why are rare gases chemically inert?
2) What is the most stable electronic state?
3) What is molar mass?
4) Which law do chemical equations have to respect?
5) What are combustion reactions?
6) What is maximum oxidation?
7) What shows that we have a complete oxidation?
8) What is an oxidation that forms CO?

# Answers: Part 12

1) They are chemically inert because they have reached the most stable electronic state.
2) It is the saturation of their last electronic layer.
3) It is the mass of a molecule.
4) The law of conservation of mass/matter.
5) These are oxidation reactions.
6) It is a complete oxidation.
7) The production of $CO_2$.
8) It is an incomplete combustion.

# Questions: Part 13

1) What is the link between viscosity and temperature?
2) Name some factors that determine the viscosity of a molecule.
3) What are the reagents in a chemical equation?
4) What are the products in a chemical equation?
5) What do the arrows represent in a chemical equation?
6) What is a mole?
7) What is a solution?
8) What is the bulk substance of a solution?
9) What is the substance in a small amount of a solution?
10) What is a concentrate of a solution?

# Answers: Part 13

1) Viscosity increases when temperature decreases.
2) Its hydrogen bridges, its shape and size.
3) These are the initial compounds.
4) These are the final compounds.
5) The direction of the reaction.
6) A mole is a unit of quantity of matter.
7) It is a homogeneous mixture of a solvent and a solute.
8) A solvent.
9) A solute.
10) This is the amount of solute dissolved in a volume of a solution.

# Questions: Part 14

1) What are the different ways of expressing concentration?
2) What is Boltzmann's distribution?
3) What is the relationship between pressure and collision?
4) What is a catalyst?
5) Is a catalyst consumed during a chemical reaction?
6) Which substances have the opposite effects of a catalyst?
7) Name a catalyst in the human body.
8) What determines an element?
9) What is the international system of units of measurement?
10) What does candela measure?

## Answers: Part 14

1) In millilitres per litre, in grams per litre, in moles per litre, in degrees and percentages.
2) It is the velocity distribution of the molecules of a gas.
3) The higher the pressure, the greater the chance of collision.
4) A catalyst triggers a chemical reaction. It increases the speed of a chemical reaction.
5) No.
6) Inhibitors.
7) An enzyme.
8) The number of protons.
9) Candela, meter, kilogram, amp, kelvin, mole and second.
10) The luminous intensity.

# Questions: Part 15

1) Can energy be created?
2) Can energy be destroyed?
3) What is potential energy?
4) How can energy be transferred?
5) Is water a polar or non-polar molecule?
6) Is oil a polar or non-polar molecule?
7) Is air a solution?
8) What does H represent in pH?
9) What is cohesion in a liquid?
10) What are the causes of cohesion in a liquid?

# Answers: Part 15

1) No.
2) No.
3) It is the energy stored in a system.
4) By work and heat.
5) Water is a polar molecule.
6) Oil is a non-polar molecule.
7) Yes.
8) Hydrogen.
9) This is the attraction between the molecules in a liquid.
10) Intermolecular forces.

# Questions: Part 16

1) What is used to measure electric current?
2) What is stoichiometry?
3) Give an example of a base substance used in our daily lives.
4) Is water an acid?
5) Is water a base?
6) What happens when an acid is added to water?
7) Define an oxidation reaction.
8) Define a reduction reaction.
9) What is entropy?
10) What kind of substance reduces the activation energy in a chemical reaction?

# Answers: Part 16

1) Ampere.
2) It is the measurement of the chemicals in a given reaction.
3) A soap.
4) Yes.
5) Yes.
6) It dissociates to form H3O+ and an anion.
7) It is the loss of electrons.
8) This is the gain of electrons.
9) It is a measure of molecular disorder or its randomness.
10) Catalysts.

# Questions: Part 17

1) When does a shift in a chemical reaction move to the right?
2) When does a shift in a chemical reaction move to the left?
3) What kind of solutions resist to changes in pH?
4) What are alkenes?
5) What are alkynes?
6) Which gas filters the ultraviolet lights from the sun?
7) What is the hardest form of carbon?
8) What is the substance that can cut a diamond?
9) Which element uses the abbreviation Hg?
10) Name the only metal in liquid form at room temperature.

# Answers: Part 17

1) When more products are formed.
2) When more reagents are formed.
3) A buffer solution.
4) Alkenes are hydrocarbons with double bonds.
5) Alkynes are hydrocarbons with triple bonds.
6) Ozone.
7) Diamond.
8) A diamond.
9) Mercury.
10) Mercury.

# Conclusion

Thank you once again for purchasing this book. I hope it has been useful to you. Please leave a review and keep on studying.

www.ingramcontent.com/pod-product-compliance
Lightning Source LLC
Chambersburg PA
CBHW031556210526
45464CB00003B/1317